Teach Your Kids Math: Adding on Your Fingers

Copyright & Other Notices

Published in 2018 by Answers 2000 Limited

Copyright © 2018, Sunil Tanna

Sunil Tanna has asserted his right to be identified as the author of this Work in accordance with the Copyright, Designs, and Patents Act 1988.

Information in this book is the opinion of the author, and is correct to the best of the author's knowledge, but is provided "as is" and without warranty to the maximum extent permissible under law. Content within this book is **not** intended as legal, tax, financial, medical, or any other form of professional advice.

While we have checked the content of this book carefully, in any educational book there is always the possibility of typographical errors, or other errors or omissions. We apologize if any such errors are found, and would appreciate if readers inform of any errors they might find, so we can update future editions/updates of this book.

Answers 2000 Limited is a private limited company registered in England under company number 3574155. Address and other information about information about Answers 2000 Limited can be found at www.ans2000.com

Updates, news & related resources from the author can be found at
http://www.suniltanna.com/addition

Information about other math books by the same author can be found at
http://www.suniltanna.com/math

Introduction

For some time now, I have tutored both children and adults in math and science. This book is based on my own personal experience as a tutor, and is one of a series of books on different topics:

- If you want to find out about the other math books that I have written, please visit: http://www.suniltanna.com/math

- For science books that I have written, visit: http://www.suniltanna.com/science

Required Skills Before Learning Adding

This book is about learning to add on your fingers. Some very basic math skills are required before starting to learn to add:

(1) Understanding the idea of what a number is.

(2) Know what 0 is.

(3) Being able to count – preferably from 0 to over 20. If your child can count to 100 (or more), even better. But they can improve counting and learn adding at the same time – so help your child with counting to higher numbers as they progress through this book.

(4) Recognize the numbers when written down – it will be hard to make good use of the pictures if they can't.

The Structure of This Book:

This book is intended as a guide to help with parents and teachers teaching addition to children, although of course it can be used by adult learners too.

- The introduction to this book, which you are reading now, is written for the parent or teacher.
- The conclusion of this book (at the end) is also written for the parent or teacher.
- Just before the conclusion are 50 questions and answers. You can work through these questions with your child, or invent your own questions, or do both.

- The rest of the book is written for the child and addresses the child directly as "you". I recommend that the parent or teacher read aloud from each chapter to the child – demonstrating the examples, and helping the child do the same on their fingers. Don't be afraid to invent your own examples while working through the book. During this process, take regular breaks to practice.

The Best Way to Learn:

As far as mastering addition is concerned, the best way to do this:

(1) Practice often and regularly

If you want to get better at something, you need to practice. The same principle applies to your child learning and mastering a new skill. Regular practice is vital!

(2) You can make up your own problems

Once you have exhausted the problems in this book, that doesn't mean you should stop! Feel free to invent your own problems. You know your child better than me, so you can tailor the problems to their current skill level and to the areas that they need to practice.

(3) Do lots of activities involving addition

It's important to devote time to learning, but you don't need to make it boring. Young children learn better when learning is fun! You can make addition part of your daily routine – ask your child to do a problem after meals or before story-time at night. Play games involving adding up numbers, play an adding quiz when going on a car ride, etc.

I have placed some more information about some resources for making adding fun, including links, at http://www.suniltanna.com/addition

(4) Go slowly and allow your kids time

I mean this in three senses:

Firstly, there is no need to rush through the concepts and the examples. Spend as much time on each idea as necessary before moving on. Once your child understands a concept and can work-out the right answer for themselves, then move forward.

Secondly, it takes time to master a new skill: you don't have to complete the whole book in a day. Doing a little bit every day and allowing your child time to digest the concepts is much more effective than trying to sprint your way through the book.

Thirdly, when your child is solving a problem, give them time to solve it, and let them work at their own pace. **Don't rush them!**
- The most important thing is that they know **how** to work out the answer.
- The second most important thing is that they get the right answer.
- The least important thing is speed – speed will come naturally with regular practice.

Rushing your child can damage their confidence, make the activity less fun (making it harder to engage them in future lessons), and tends to increase the number of mistakes (which further damages their confidence).

That said it is perfectly okay to help your child when doing problems. Eventually of course you want them to be able to do problems by themselves, but when they are starting out that may be a little too much to ask. The way that I do this when tutoring is:

- I show the child how to do an example problem.
- I ask the child to do a similar problem while watching them carefully. If it is clear that the child has **not** understood or does **not** know where to start, I will show them again. If they make a minor mistake, I might ask whether they are sure, ask them to count on my fingers instead of theirs, or simply ask them to try again.
- If you help your child in this way while still allowing to solve the problem themselves, it shouldn't be long before they learn to solve such problems without any help from you.

How to Add on Your Fingers

Suppose you want to add two numbers together, how can you do it on your fingers?

Let's try 8 + 4

(1) Say the 1st number in the sum. Stretch out the number of fingers that are in the 2nd number.

So, if you were doing 8 + 4, you would say "8" and stretch out 4 fingers like this:

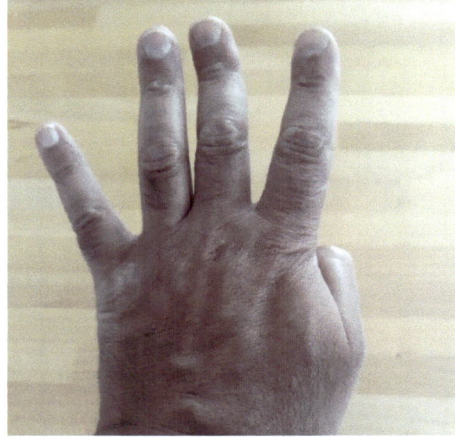

(2) Using your fingers, start counting upwards from the number you said, so the first finger is 1 more than this number, the second finger is 2 more, and so on.

If you were doing 8 + 4, you would count upwards from 8 like this:

You would have already said "8", and then your fingers count-up "9, 10, 11, 12":

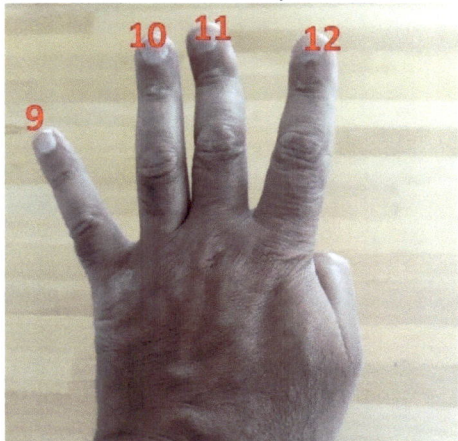

(3) The last number that you say is the answer! So, 8 + 4 = 12.

Let's try 3 + 6

(1) First you say "3". Then you stretch out 6 fingers:

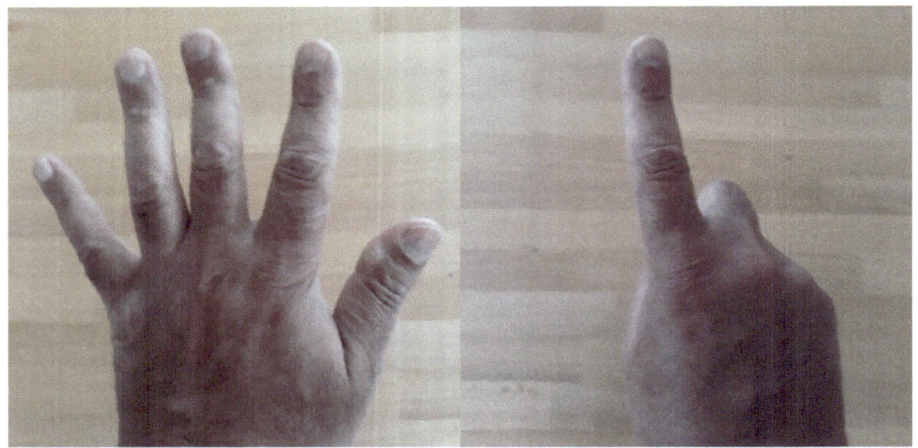

(2) You count upwards from 3.

Your fingers count "4, 5, 6, 7, 8…" and then "9":

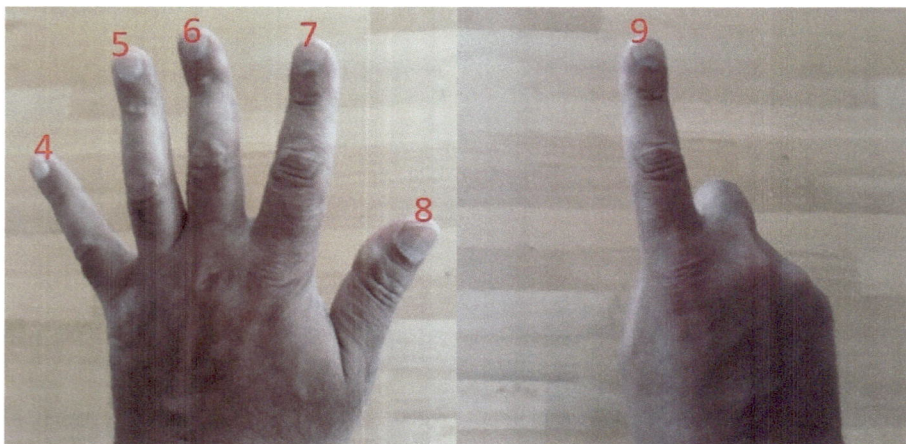

(3) The last number was 9. So, 3 + 6 = 9.

Let's try 4 + 5

(1) First you say "4". Then you stretch out 5 fingers:

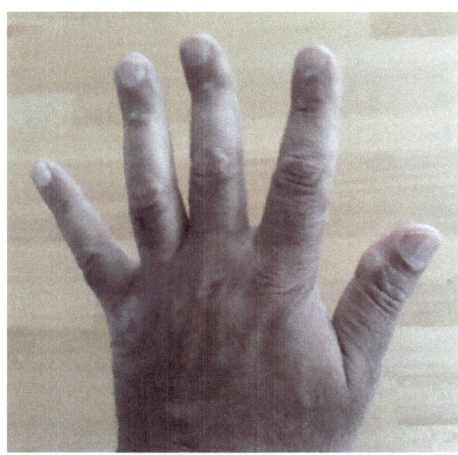

(2) You count upwards from 4.

Your fingers count "5, 6, 7, 8, 9":

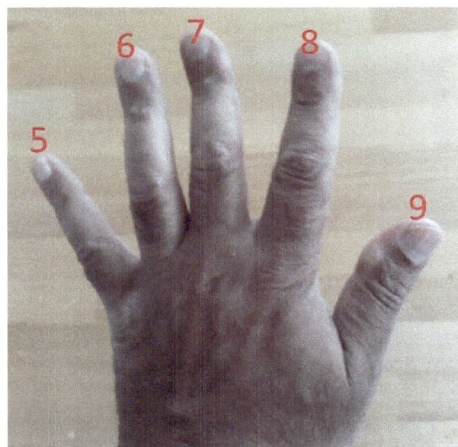

(3) The last number was 9. So, 4 + 5 = 9.

Let's try 5 + 7

(1) First you say "5". Then you stretch out 7 fingers:

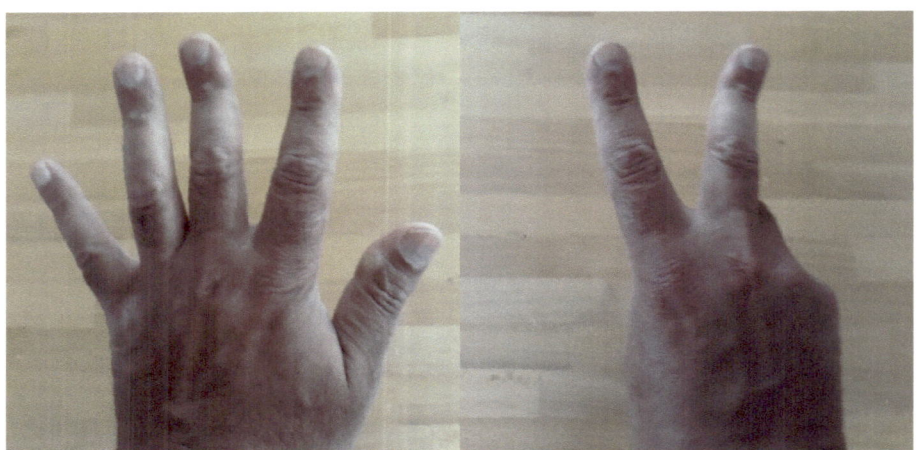

(2) You count upwards from 5.

Your fingers count "6, 7, 8, 9, 10…" and then "11, 12":

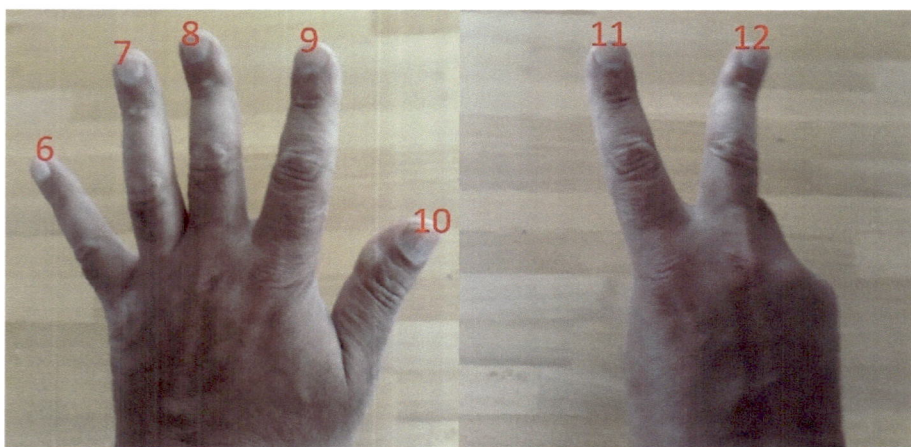

(3) The last number was 12. So, 5 + 7 = 12.

Teach Your Kids Math: Adding on Your Fingers

Let's try 9 + 3

(1) First you say "9". Then you stretch out 3 fingers:

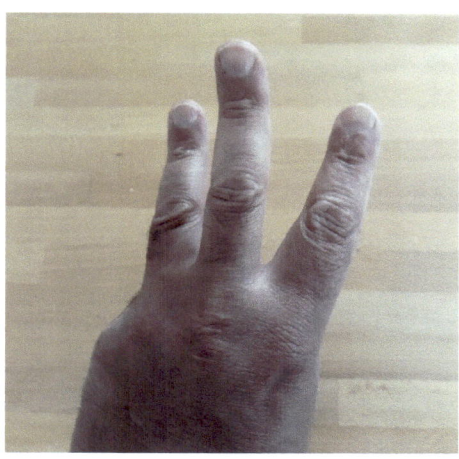

(2) You count upwards from 9.

Your fingers count "10, 11, 12":

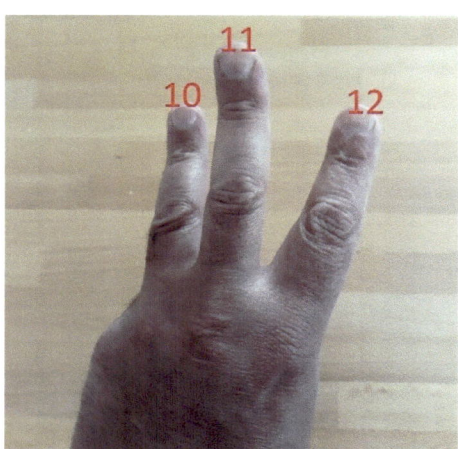

(3) The last number was 12. So, 9 + 3 = 12.

Let's try 9 + 8

(1) First you say "9". Then you stretch out 8 fingers:

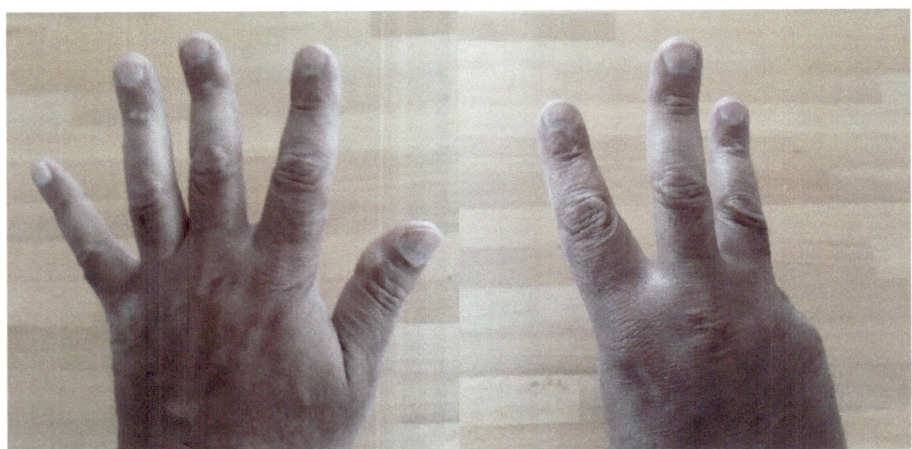

(2) You count upwards from 9.

Your fingers count "10, 11, 12, 13, 14..." and then "15, 16, 17":

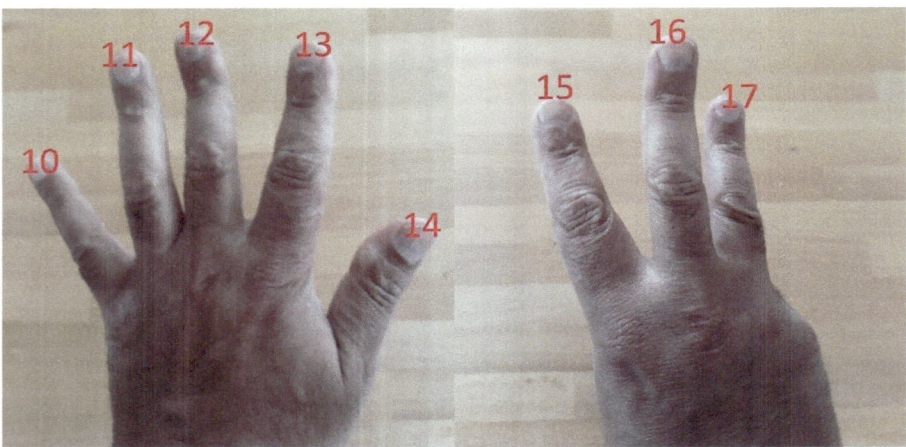

(3) The last number was 17. So, 9 + 8 = 17.

Let's try 2 + 9:

(1) First you say "2". Then you stretch out 9 fingers:

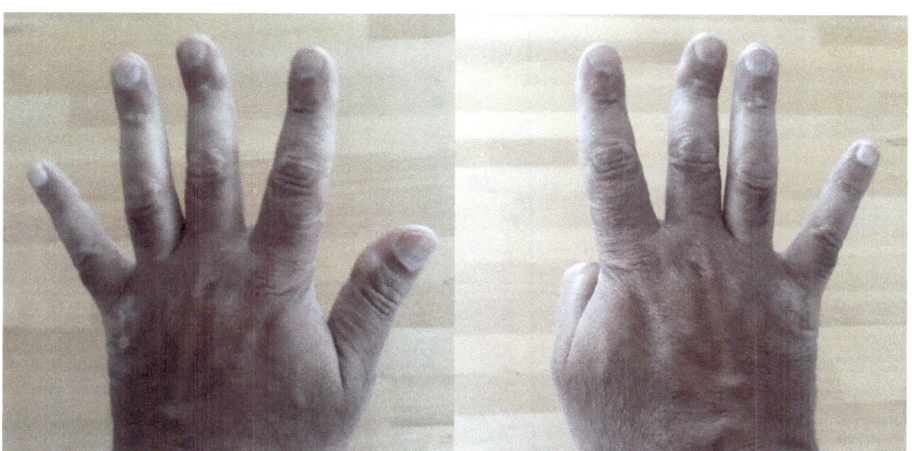

(2) You count upwards from 2.

Your fingers count "3, 4, 5, 6, 7…" and then "8, 9, 10, 11":

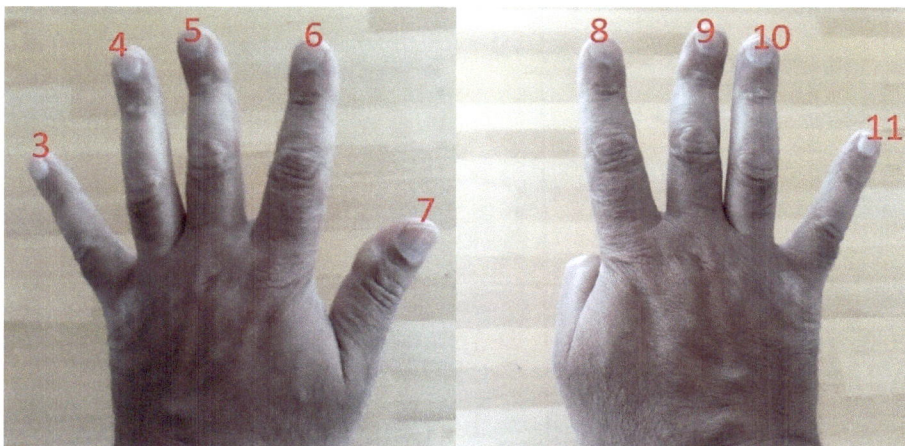

(3) The last number was 11. So, 2 + 9 = 11.

Let's try 4 + 6

(1) First you say "4". Then you stretch out 6 fingers:

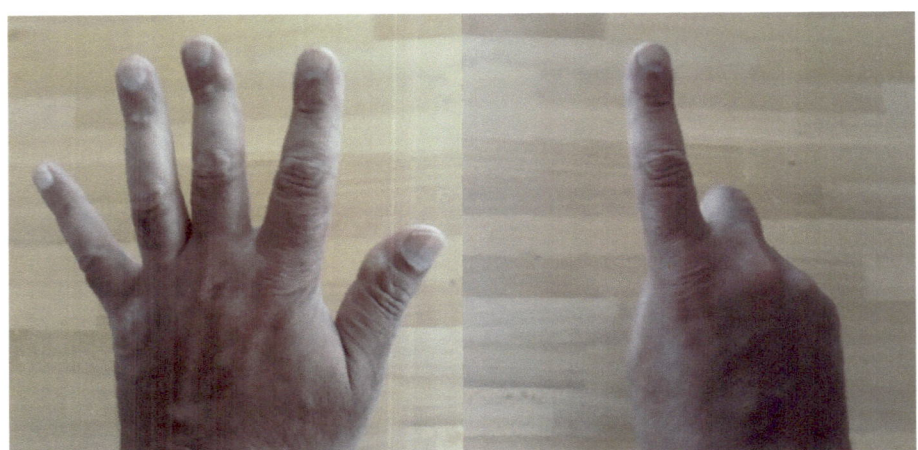

(2) You count upwards from 4.

Your fingers count "5, 6, 7, 8, 9…" and then "10":

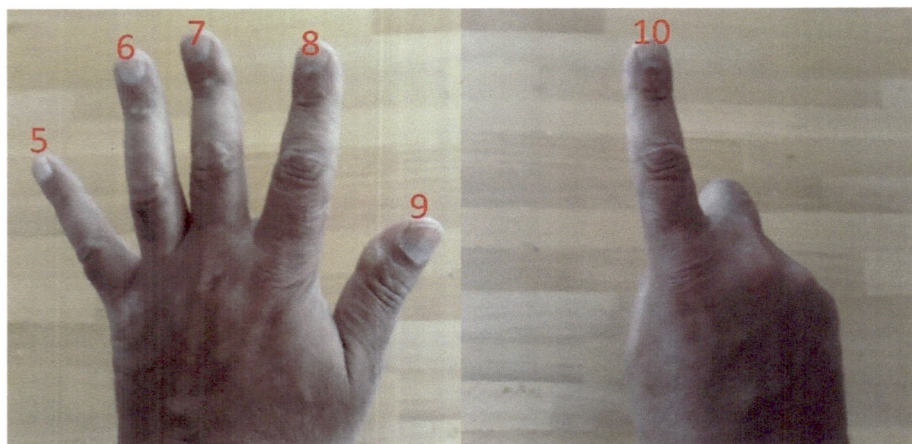

(3) The last number was 10. So, 4 + 6 = 10.

Teach Your Kids Math: Adding on Your Fingers

Do you want to try some bigger numbers?

Now let's have a go with some questions where the answer is over 20.

Let's try 18 + 3

(1) First you say "18". Then you stretch out 3 fingers:

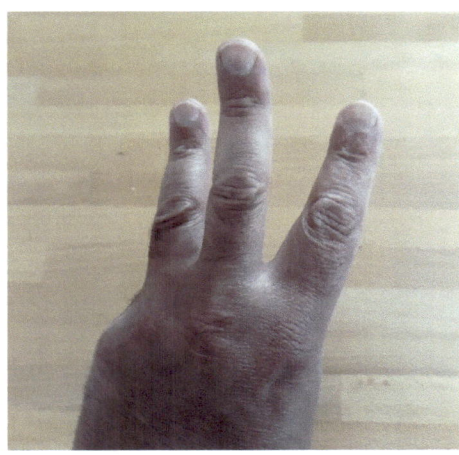

(2) You count upwards from 18.

Your fingers count-up "19, 20, 21":

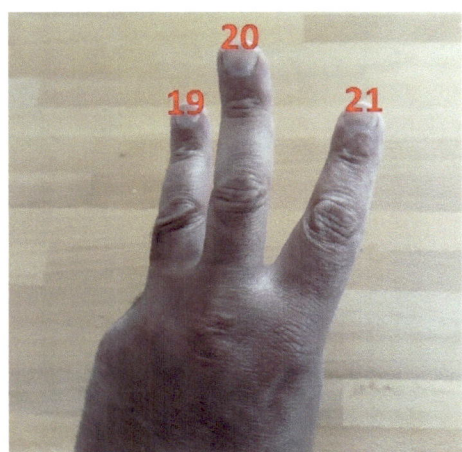

(3) The last number was 21. So, 18 + 3 = 21.

Teach Your Kids Math: Adding on Your Fingers

Let's try 26 + 8

(1) First you say "26". Then you stretch out 8 fingers:

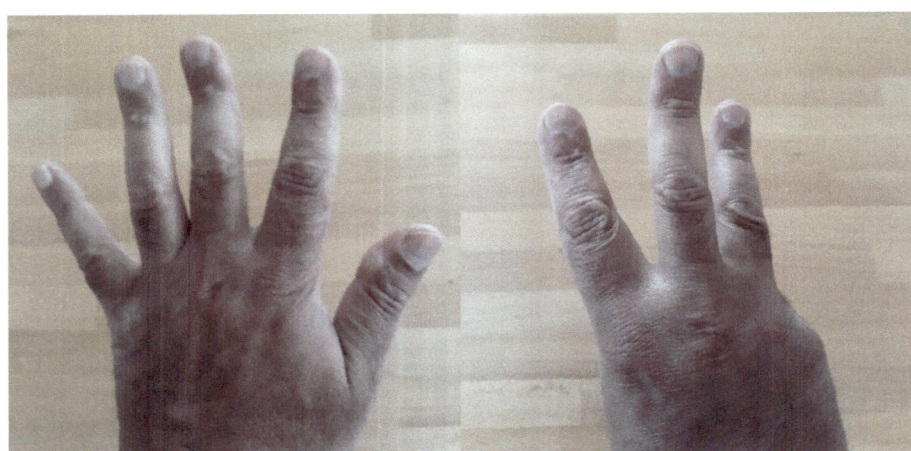

(2) You count upwards from 26.

Your fingers count "27, 28, 29, 30, 31…" and then "32, 33, 34":

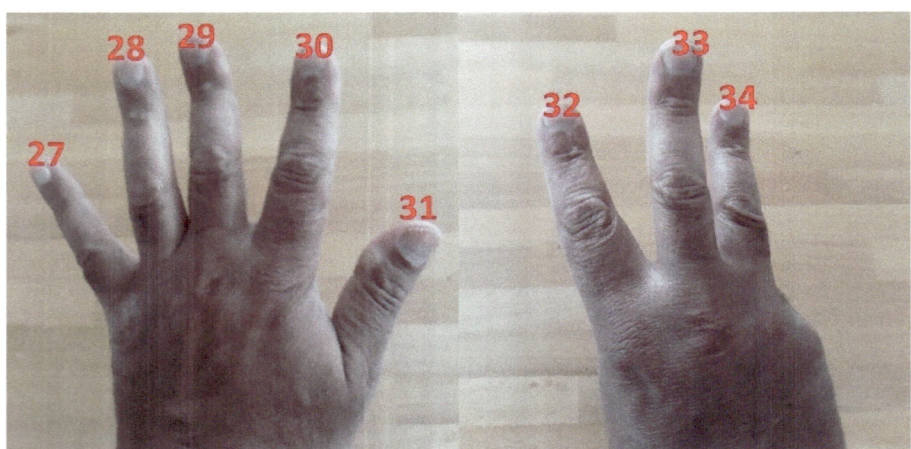

(3) The last number was 34. So, 26 + 8 = 34.

Let's try 37 + 5

(1) First you say "37". Then you stretch out 5 fingers:

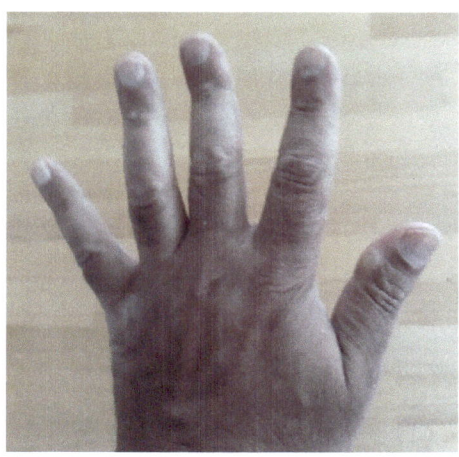

(2) You count upwards from 37.

Your fingers count "38, 39, 40, 41, 42":

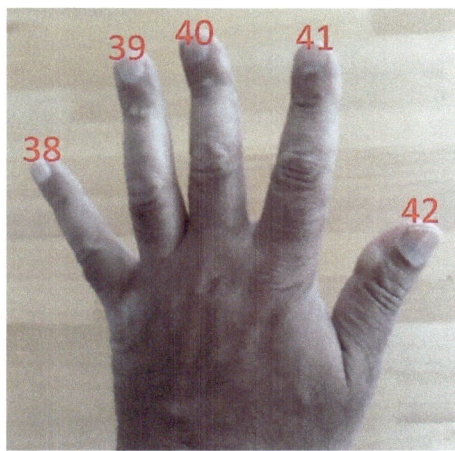

(3) The last number was 42. So, 37 + 5 = 42.

Let's try 22 + 8

(1) First you say "22". Then you stretch out 8 fingers:

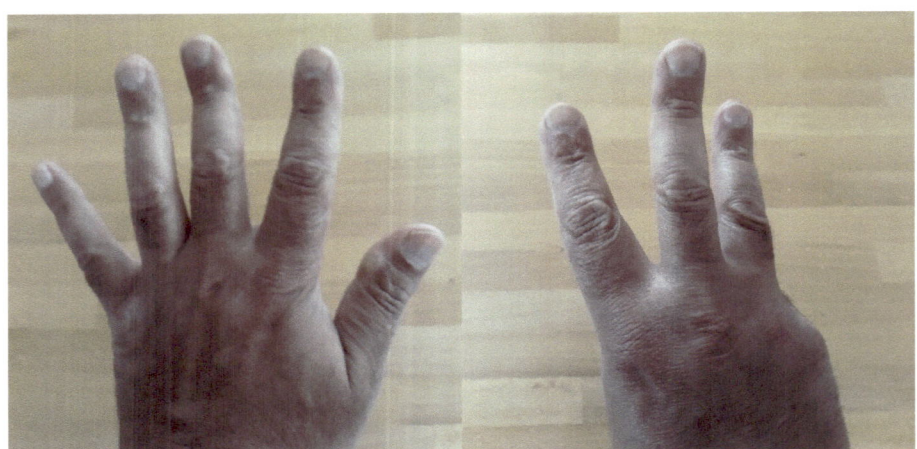

(2) You count upwards from 22.

Your fingers count "23, 24, 25, 26, 27…" and then "28, 29, 30":

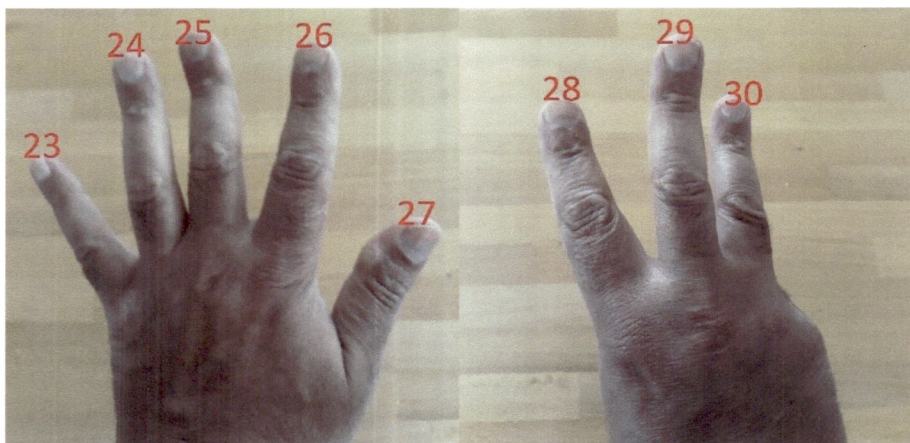

(3) The last number was 30. So, 22 + 8 = 30.

Let's try 37 + 2

(1) First you say "37". Then you stretch out 2 fingers:

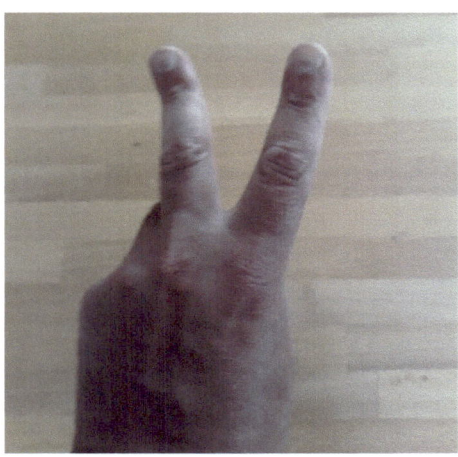

(2) You count upwards from 37.

Your fingers count "38, 39":

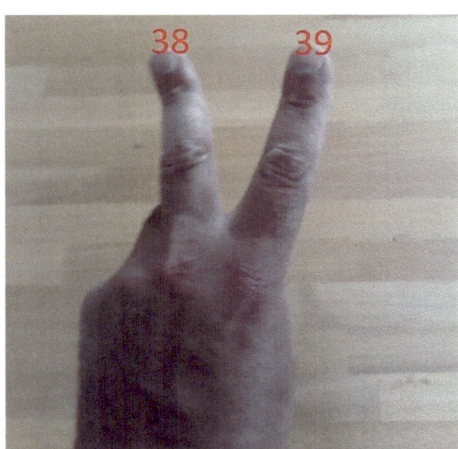

(3) The last number was 39. So, 37 + 2 = 39.

Let's try 25 + 9

(1) First you say "25". Then you stretch out 9 fingers:

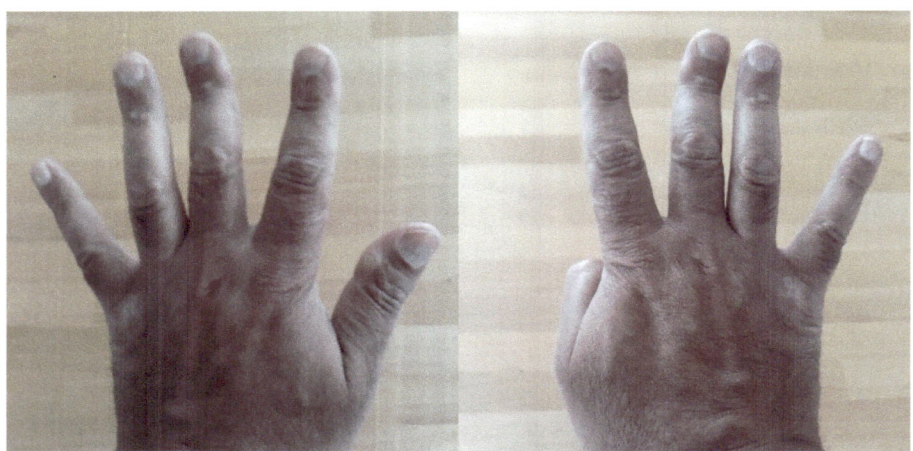

(2) You count upwards from 25.

Your fingers count "26, 27, 28, 29, 30…" and then "31, 32, 33, 34":

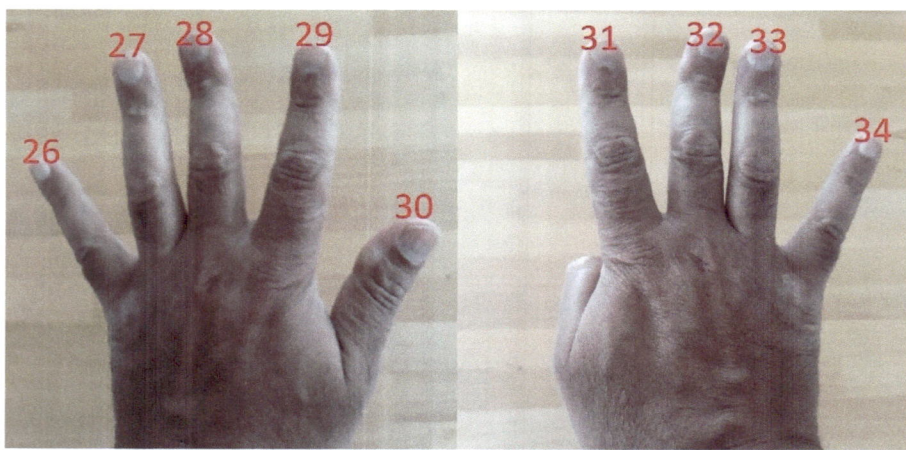

(3) The last number was 34. So, 25 + 9 = 34.

Let's try 15 + 8

(1) First you say "15". Then you stretch out 8 fingers:

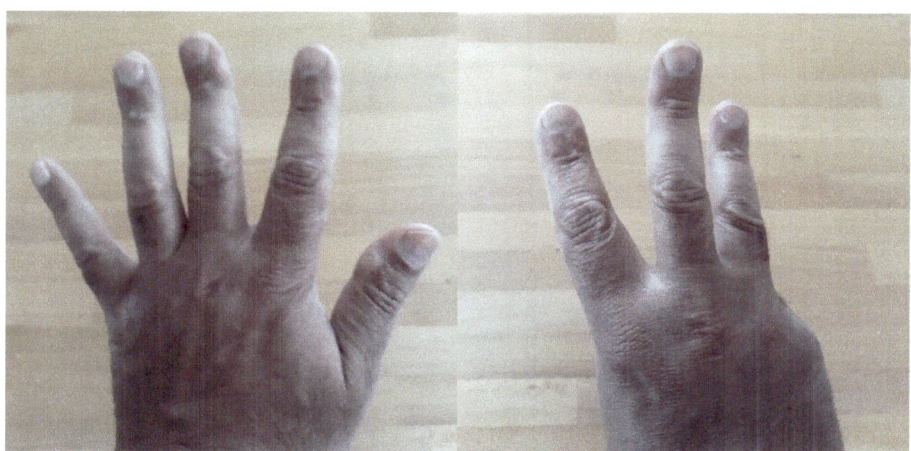

(2) You count upwards from 15.

Your fingers count "16, 17, 18, 19, 20..." and then "21, 22, 23":

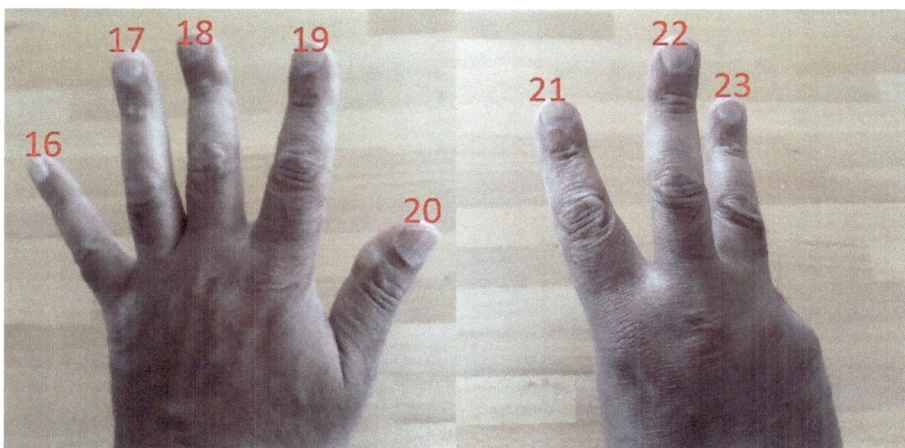

(3) The last number was 23. So, 15 + 8 = 23.

What about if wanted to add more than two numbers?

Let's try 9 + 8 + 7

We do it exactly like before, but we do it slowly step-by-step.

(1) You are going to start with 9 and add 8.

So, you say "9" and stretch out 8 fingers:

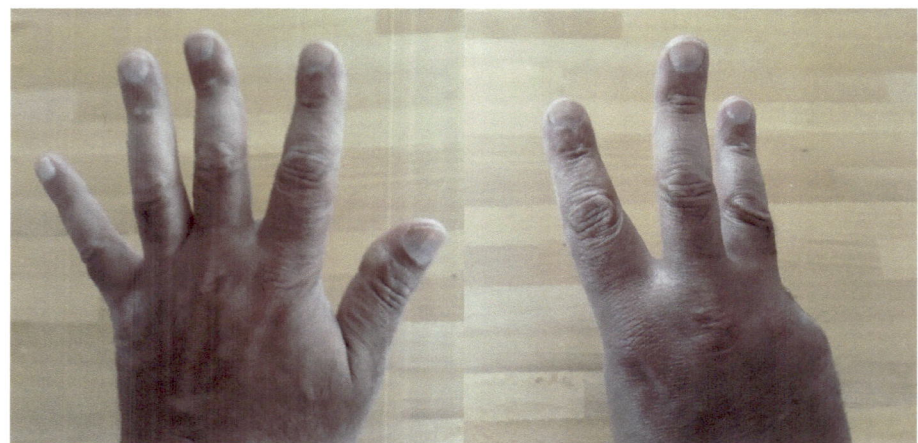

(2) Now you count upwards from 9.

Your fingers count "10, 11, 12, 13, 14…" and then "15, 16, 17":

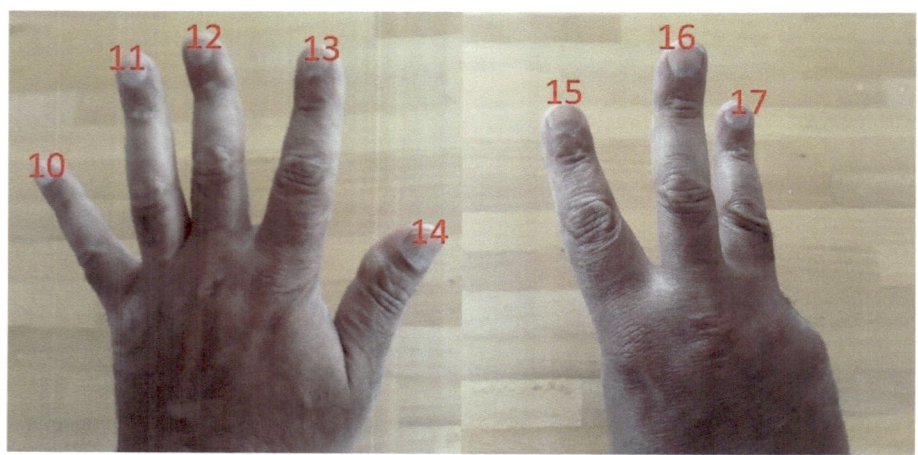

(3) Since the last number was 17 that means 9 + 8 = 17. But you're **not** finished yet! You still need to add the 7!

(4) So, you stretch out 7 fingers:

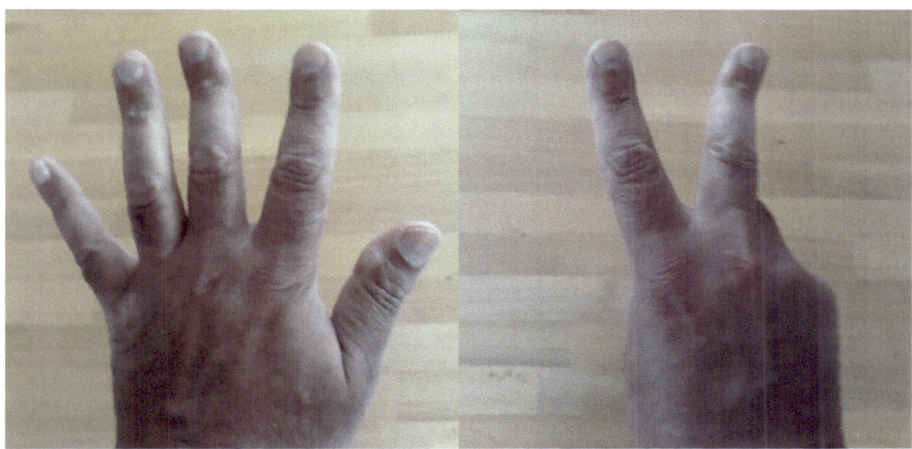

(5) And you keep on counting upwards.

Your fingers count "18, 19, 20, 21, 22…" and then "23, 24":

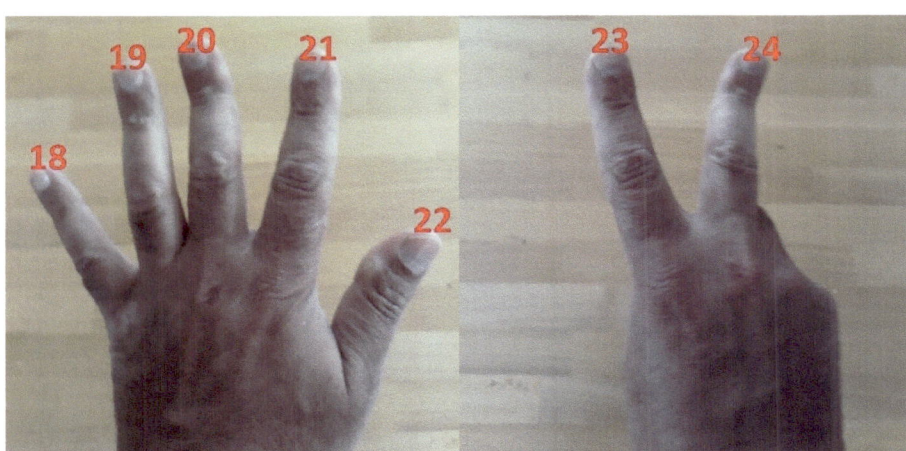

(6) Since the last number was 24 and you've used all the numbers that means 9 + 8 + 7 = 24.

Let's try 13 + 4 + 8

(1) You are going to start with 13 and add 4.

So, you say "13" and stretch out 4 fingers:

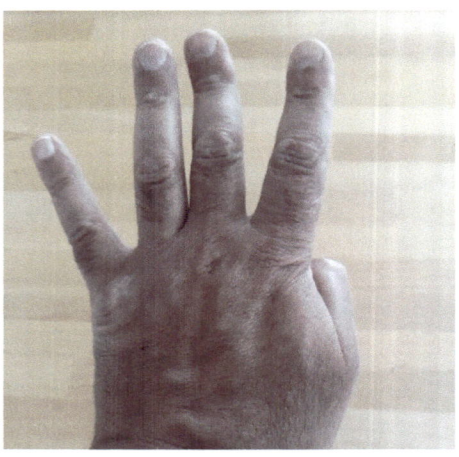

(2) Now you count upwards from 13.

Your fingers count "14, 15, 16, 17":

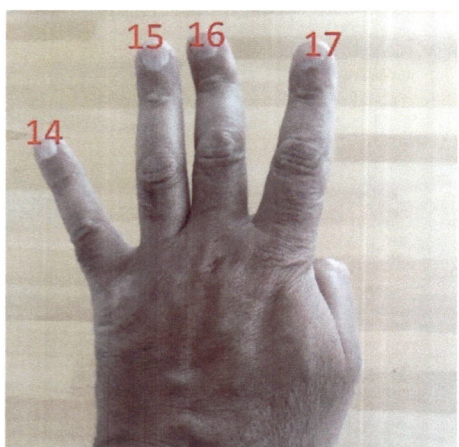

(3) Since the last number was 17 that means 13 + 4 = 17. But you're **not** finished yet! You still need to add the 8!

(4) So, you stretch out 8 fingers:

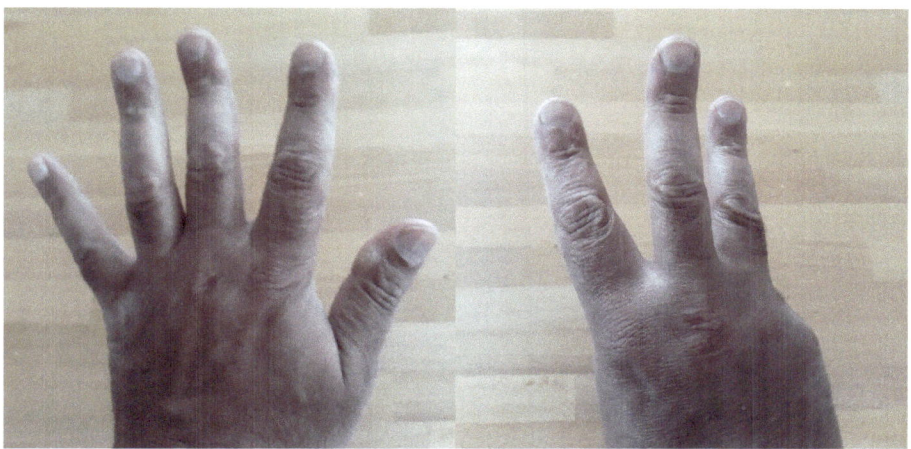

(5) And you keep on counting upwards.

Your fingers count "18, 19, 20, 21, 22…" and then "23, 24, 25":

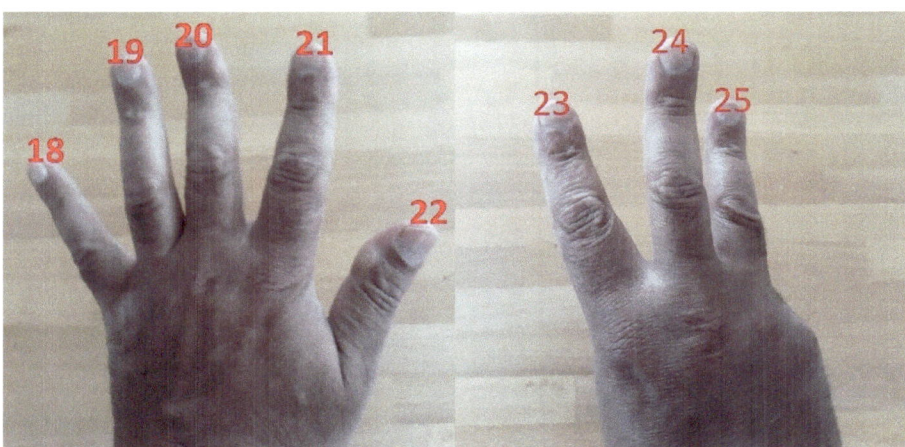

(6) Since the last number was 25 and you've used all the numbers that means 13 + 4 + 8 = 25.

Let's try 19 + 6 + 3

(1) You are going to start with 19 and add 6.

So, you say "19" and stretch out 6 fingers:

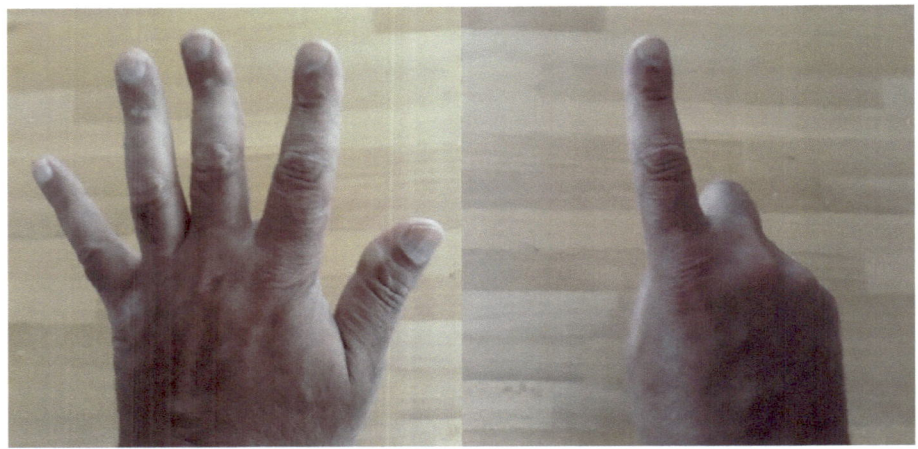

(2) Now you count upwards from 19.

Your fingers count "20, 21, 22, 23, 24…" and then "25":

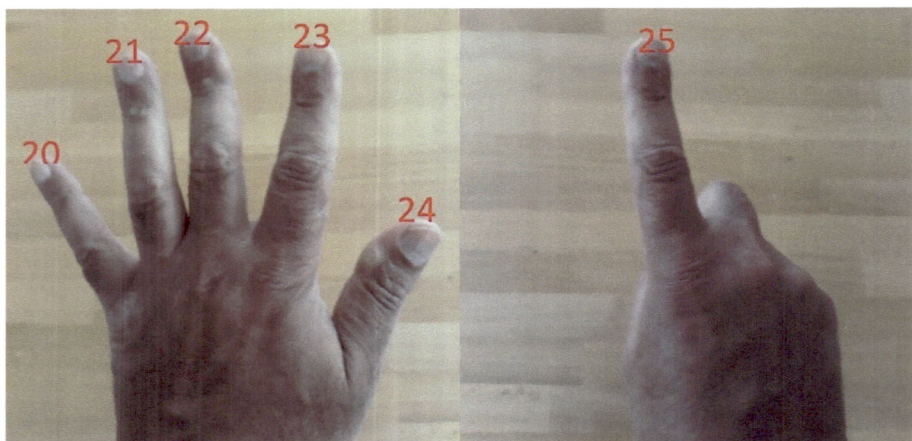

(3) Since the last number was 25 that means 19 + 6 = 25. But you're **not** finished yet! You still need to add the 3!

(4) So, you stretch out 3 fingers:

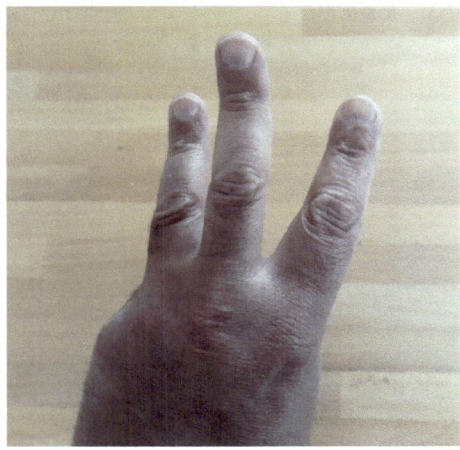

(5) And you keep on counting upwards.

Your fingers count "26, 27, 28":

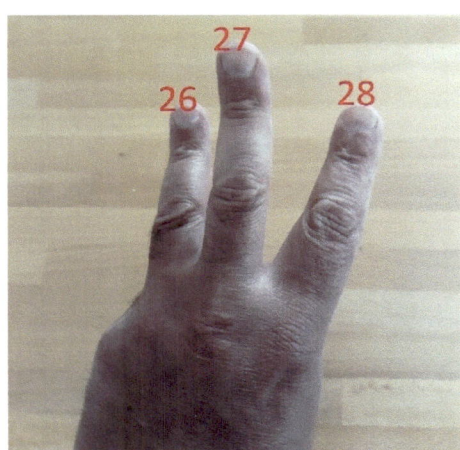

(6) Since the last number was 28 and you've used all the numbers that means 19 + 6 + 3 = 28.

Let's try 11 + 9 + 3

(1) You are going to start with 11 and add 9.

So, you say "11" and stretch out 9 fingers:

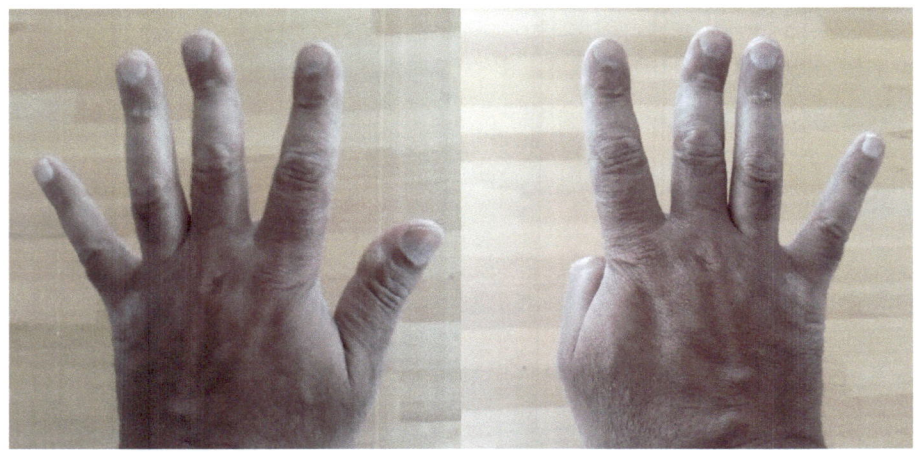

(2) Now you count upwards from 11.

Your fingers count "12, 13, 14, 15, 16…" and then "17, 18, 19, 20":

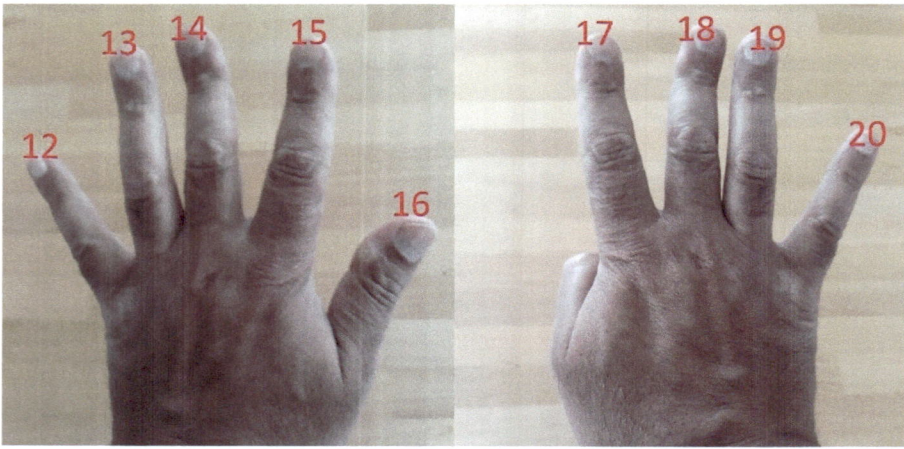

(3) Since the last number was 20 that means 11 + 9 = 20. But you're **not** finished yet! You still need to add the 3!

(4) So, you stretch out 3 fingers:

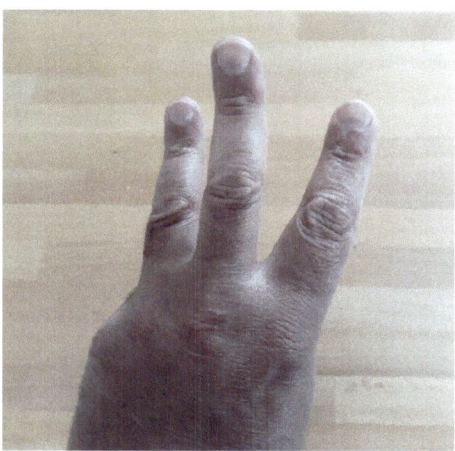

(5) And you keep on counting upwards.

Your fingers count "21, 22, 23":

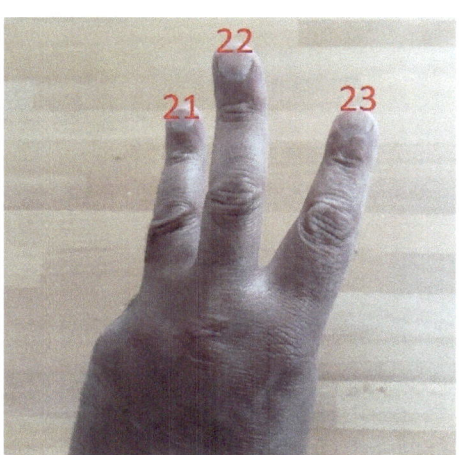

(6) Since the last number was 23 and you've used all the numbers that means 11 + 9 + 3 = 23.

Teach Your Kids Math: Adding on Your Fingers

Let's try 4 + 8 + 5 + 7

Now we're adding four numbers!

This might seem more difficult, but it isn't. We just need to go carefully step-by-step.

(1) You are going to start with 4 and add 8.

So, you say "4" and stretch out 8 fingers:

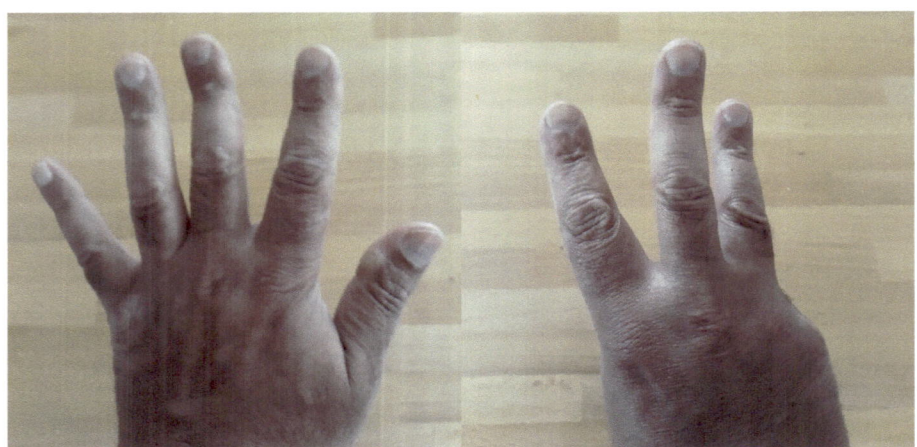

(2) Now you count upwards from 4.

Your fingers count "5, 6, 7, 8, 9…" and then "10, 11, 12":

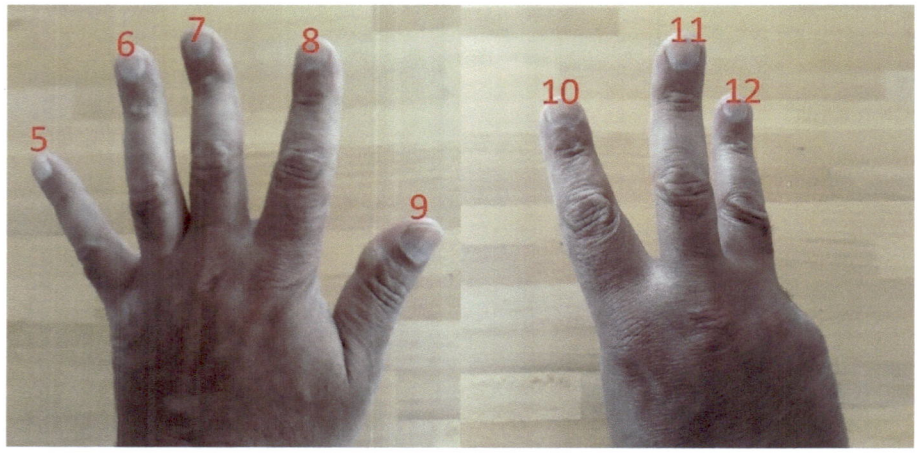

(3) Since the last number was 12 that means 4 + 8 = 12. But you're **not** finished yet! You still need to add the 5 and the 7!

(4) So, you stretch out 5 fingers:

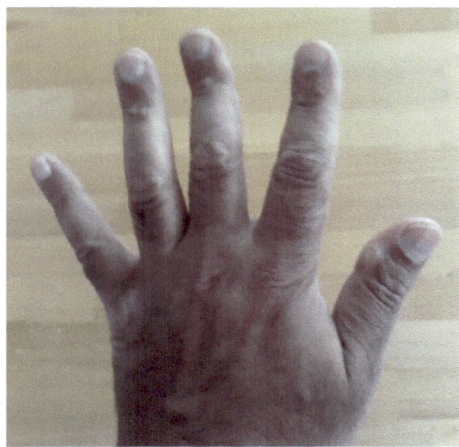

(5) And you keep on counting upwards.

Your fingers count "13, 14, 15, 16, 17":

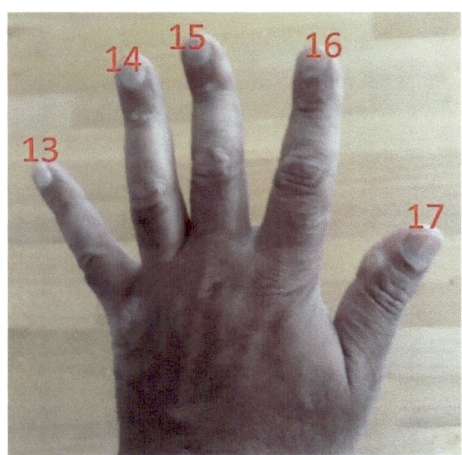

(6) Since the last number was 17 that means 4 + 8 + 5 = 17. But you're still **not** finished yet! You still need to add the 7!

Just one more number to add…

(7) So, you stretch out 7 fingers:

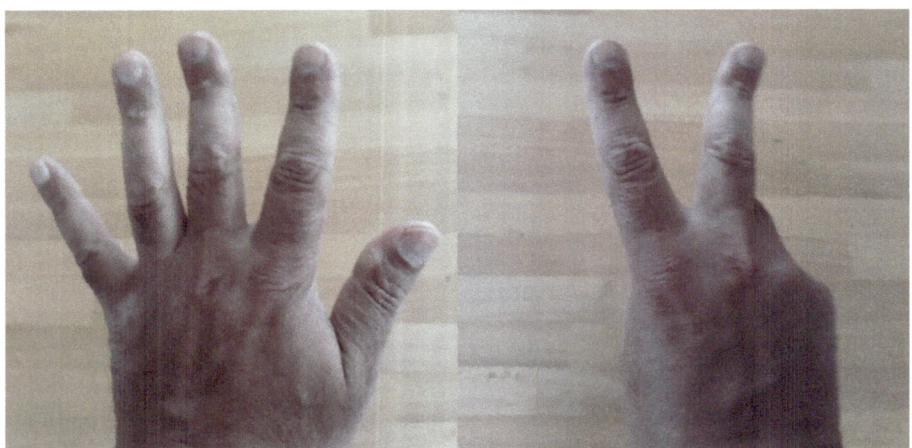

(8) And you keep on counting upwards.

Your fingers count "18, 19, 20, 21, 22…" and then "23, 24":

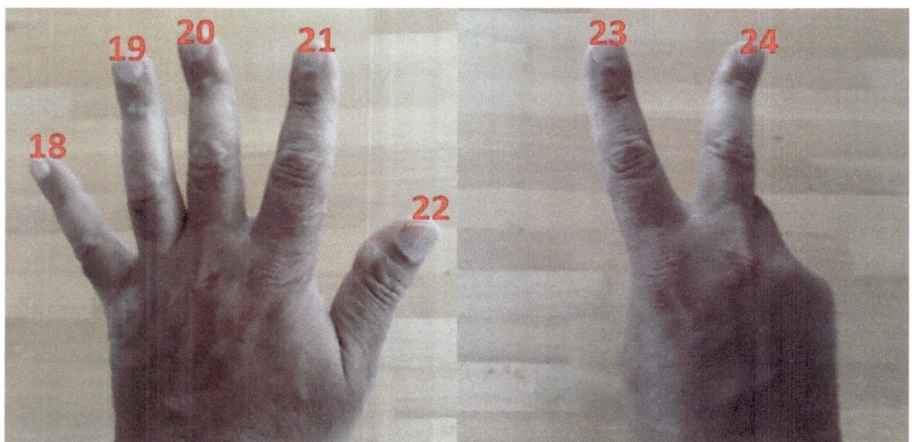

(9) Since the last number was 24 and you've used all the numbers that means 4 + 8 + 5 + 7 = 24.

Let's try 16 + 5 + 6 + 9

Once again, we go carefully step-by-step.

(1) You are going to start with 16 and add 5.

So, you say "16" and stretch out 5 fingers:

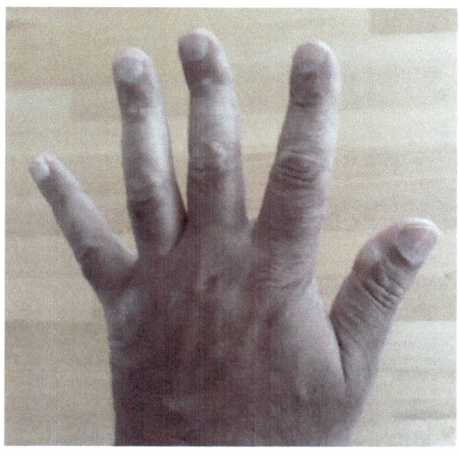

(2) Now you count upwards from 16.

Your fingers count "17, 18, 19, 20, 21":

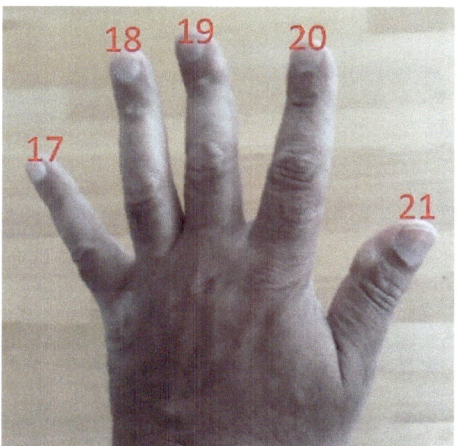

(3) Since the last number was 21 that means 16 + 5 = 21. But you're **not** finished yet! You still need to add the 6 and the 9!

(4) So, you stretch out 6 fingers:

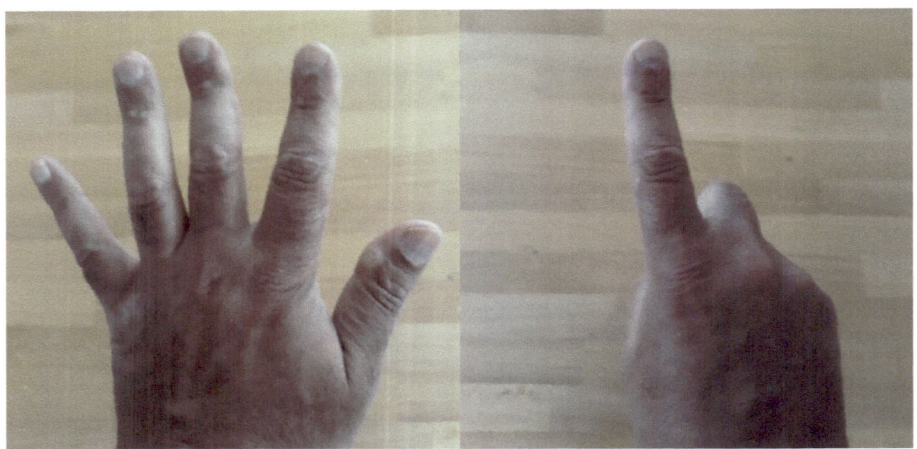

(5) And you keep on counting upwards.

Your fingers count "22, 23, 24, 25, 26…" and then "27":

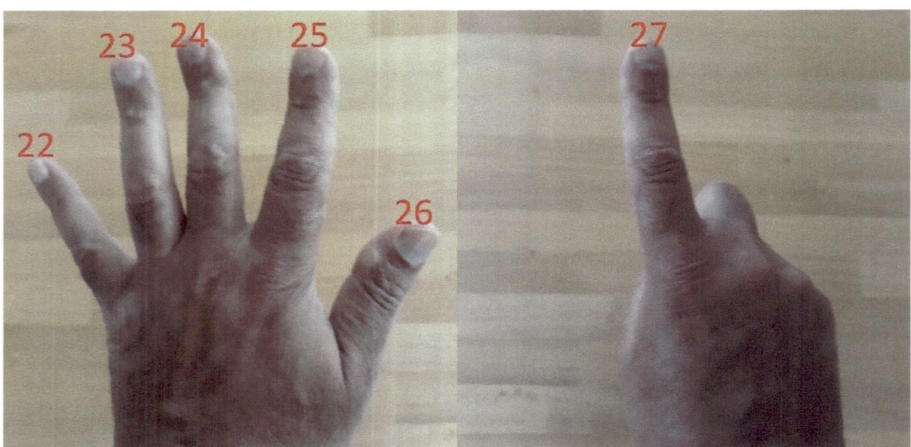

(6) Since the last number was 27 that means 16 + 5 + 6 = 27. But you're still **not** finished yet! You still need to add the 9!

Just one more number to add…

(7) So, you stretch out 9 fingers:

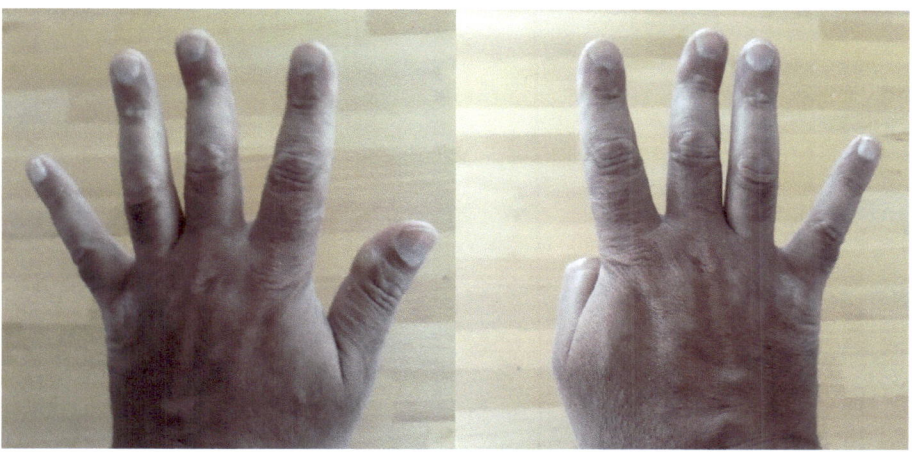

(8) And you keep on counting upwards.

Your fingers count "28, 29, 30, 31, 32…" and then "33, 34, 35, 36":

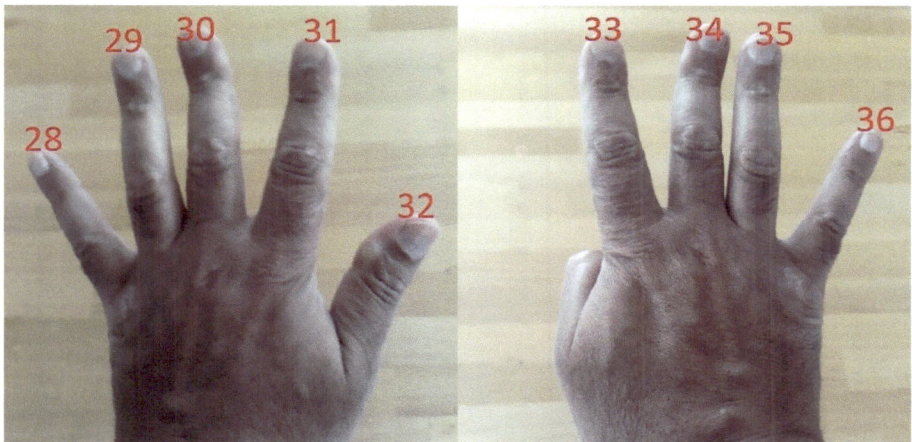

(9) Since the last number was 36 and you've used all the numbers that means 16 + 5 + 6 + 9 = 36.

Let's try 11 + 3 + 9 + 2

Once again, we go carefully step-by-step.

(1) You are going to start with 11 and add 3.

So, you say "11" and stretch out 3 fingers:

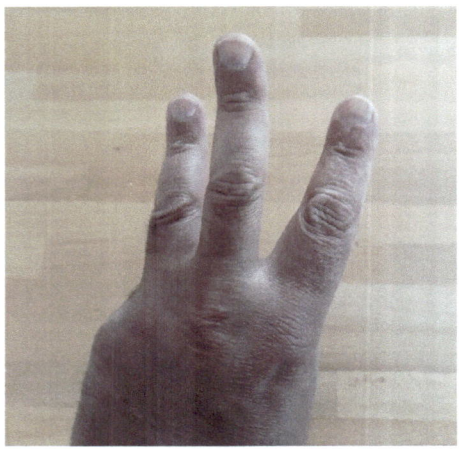

(2) Now you count upwards from 11.

Your fingers count "12, 13, 14":

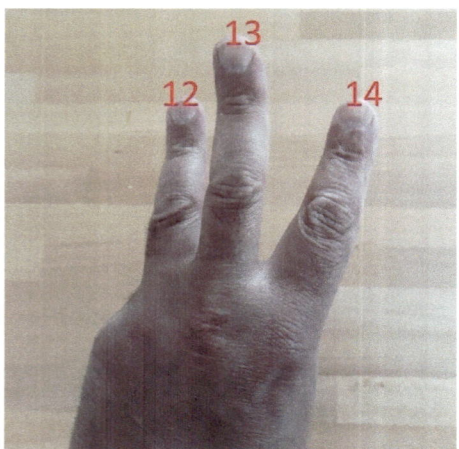

(3) Since the last number was 14 that means 11 + 3 = 14. But you're **not** finished yet! You still need to add the 9 and the 2!

(4) So, you stretch out 9 fingers:

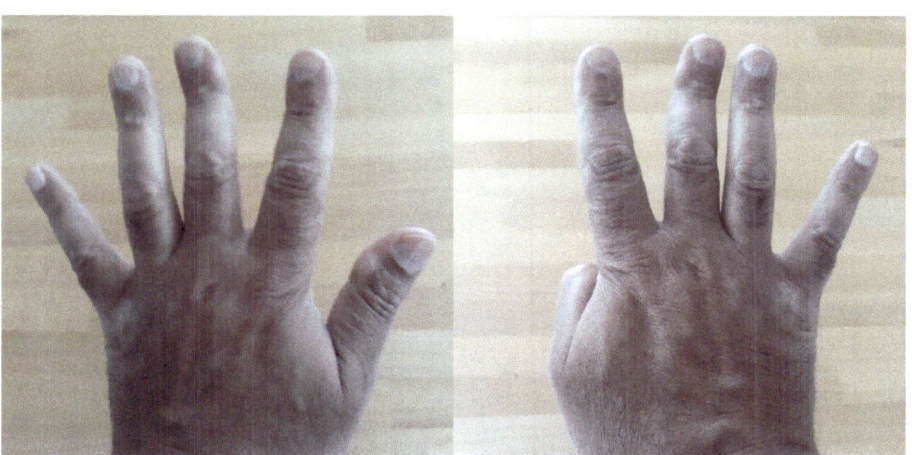

(5) And you keep on counting upwards.

Your fingers count "15, 16, 17, 18, 19…" and then "20, 21, 22, 23":

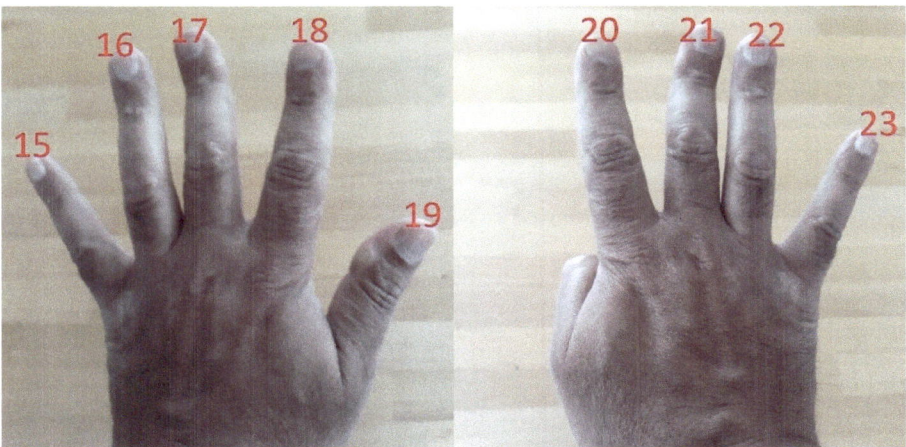

(6) Since the last number was 23 that means 11 + 3 + 9 = 23. But you're still **not** finished yet! You still need to add the 2!

Just one more number to add…

(7) So, you stretch out 2 fingers:

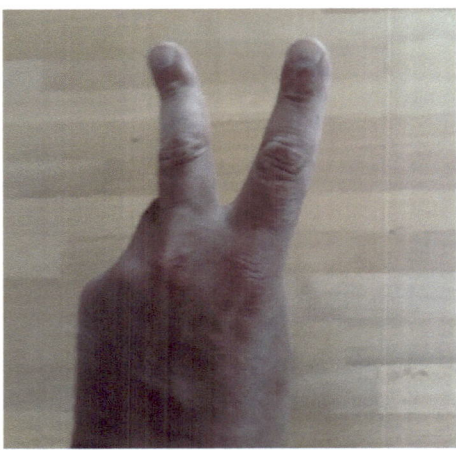

(8) And you keep on counting upwards.

Your fingers count "24, 25":

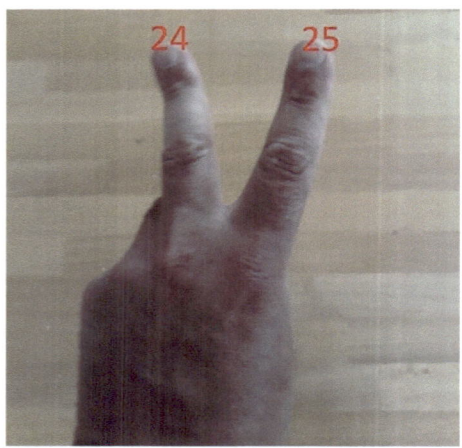

(9) Since the last number was 25 and you've used all the numbers that means 11 + 3 + 9 + 2 = 25.

One last thing!

What if you want to add 0?

If you add 0 to any number, then that number does **not** change:

- 5 + 0 = 5
- 3 + 0 = 3
- 10 + 0 = 10
- 4 + 0 = 4

Do you know what 7 + 0 is?

Yes 7!

What about 11 + 0?

Yes, it's 11.

What about 23 + 0?

Yes, it's 23.

What about 1000 + 0?

I think you know…

Yes, it's 1000.

Teach Your Kids Math: Adding on Your Fingers

Can You Answer These Questions?

Let's start with some easy questions:

1. 4 + 3?
2. 3 + 4?
3. 6 + 5?
4. 9 + 8?
5. 7 + 5?
6. 8 + 5?
7. 11 + 4?
8. 7 + 0?
9. 5 + 9?
10. 9 + 0?
11. 13 + 4?
12. 8 + 6?
13. 9 + 7?
14. 6 + 9?
15. 6 + 0?
16. 13 + 4?
17. 13 + 0?
18. 14 + 5?
19. 9 + 5?
20. 7 + 7?

Now some slightly harder questions with bigger numbers:

21. 13 + 8?
22. 24 + 9?

23. 31 + 9?

24. 27 + 4?

25. 25 + 9?

26. 47 + 5?

27. 16 + 9?

28. 36 + 7?

29. 46 + 0?

30. 25 + 7?

31. 32 + 8?

32. 32 + 0?

33. 14 + 8?

34. 24 + 8?

35. 34 + 8?

36. 44 + 8?

37. 16 + 9?

38. 26 + 9?

39. 36 + 9?

40. 49 + 6?

Finally, let's try some questions that require adding three, four or even five numbers:

41. 9 + 4 + 3?

42. 9 + 9 + 8?

43. 6 + 7 + 8?

44. 6 + 7 + 8 + 0?

45. 16 + 5 + 9?

46. 23 + 9 + 6?

47. 14 + 9 + 7 + 2?

48. 24 + 3 + 5 + 9?

49. 16 + 6 + 5 + 4 + 1?

50. 16 + 0 + 9 + 0 + 9?

Teach Your Kids Math: Adding on Your Fingers

Answers

1. 4 + 3 = 7
2. 3 + 4 = 7
3. 6 + 5 = 11
4. 9 + 8 = 17
5. 7 + 5 = 12
6. 8 + 5 = 13
7. 11 + 4 = 15
8. 7 + 0 = 7
9. 5 + 9 = 14
10. 9 + 0 = 9
11. 13 + 4 = 17
12. 8 + 6 = 14
13. 9 + 7 = 16
14. 6 + 9 = 15
15. 6 + 0 = 6
16. 13 + 4 = 17
17. 13 + 0 = 13
18. 14 + 5 = 19
19. 9 + 5 = 14
20. 7 + 7 = 14
21. 13 + 8 = 21
22. 24 + 9 = 33
23. 31 + 9 = 40
24. 27 + 4 = 31
25. 25 + 9 = 34
26. 47 + 5 = 52

Teach Your Kids Math: Adding on Your Fingers

27. 16 + 9 = 25

28. 36 + 7 = 43

29. 46 + 0 = 46

30. 25 + 7 = 32

31. 32 + 8 = 40

32. 32 + 0 = 32

33. 14 + 8 = 22

34. 24 + 8 = 32

35. 34 + 8 = 42

36. 44 + 8 = 52

37. 16 + 9 = 25

38. 26 + 9 = 33

39. 36 + 9 = 45

40. 49 + 6 = 55

41. 9 + 4 + 3 = 16

42. 9 + 9 + 8 = 26

43. 6 + 7 + 8 = 21

44. 6 + 7 + 8 + 0 = 21

45. 16 + 5 + 9 = 30

46. 23 + 9 + 6 = 38

47. 14 + 9 + 7 + 2 = 32

48. 24 + 3 + 5 + 9 = 41

49. 16 + 6 + 5 + 4 + 1 = 32

50. 16 + 0 + 9 + 0 + 9 = 34

Conclusion

Well done for finishing the book! I hope you enjoyed it.

For more addition resources, please go to:

- http://www.suniltanna.com/addition

If you are looking for the next step, why not try learning column addition? (did you notice that in this book we always added a number less than 10? With column addition you can do sums like 29 + 43). You can learn column addition using my book Teach Your Kids Math: Column Addition.

To find out about other educational books that I have written, please go to:

- For math books: http://www.suniltanna.com/math
- For science books: http://www.suniltanna.com/science

Remember: If you enjoyed this book or it helped you, please post a positive review on Amazon!

www.ingramcontent.com/pod-product-compliance
Lightning Source LLC
Chambersburg PA
CBHW040416220526
45473CB00004B/1258